LEONARD MUBALAMA

Synergy of Traditional and Modern Knowledge: Conservation Paradigm

LEONARD MUBALAMA

Synergy of Traditional and Modern Knowledge: Conservation Paradigm

Imprint
Any brand names and product names mentioned in this book are subject to trademark, brand or patent protection and are trademarks or registered trademarks of their respective holders. The use of brand names, product names, common names, trade names, product descriptions etc. even without a particular marking in this work is in no way to be construed to mean that such names may be regarded as unrestricted in respect of trademark and brand protection legislation and could thus be used by anyone.

Cover image: www.ingimage.com

This book is a translation from the original published under ISBN 978-613-8-46861-5.

Publisher:
Sciencia Scripts
is a trademark of
Dodo Books Indian Ocean Ltd., member of the OmniScriptum S.R.L Publishing group
str. A.Russo 15, of. 61, Chisinau-2068, Republic of Moldova Europe
Printed at: see last page
ISBN: 978-620-4-09786-2

Table of contents:

Traditional knowledge reconciled with scientific knowledge as a new paradigm for the conservation of Protected Areas: The case of Malambo in the Itombwe Nature Reserve, DR Congo

Mubalama Kakira Léonard1 &, Igunzi Alonda Félix2, Jean Claude Kyungu2
[1] Department of Environment and Sustainable Development, Institut Supérieur deDéveloppement Rural-Bukavu, DR.Congo
[2] Itombwe Nature Reserve, Institut Congolais pour la Conservation de la Nature, D.R. Congo.

Abstract

Traditional knowledge reconciled with the scientific knowledge as a new paradigm of the conservation of protected areas: The Itombwe Natural Reserve Malambo case study, DR Congo

The relevant question of the protection of the traditional knowledge of the indigenous people is at the crossing cutting edge of several big challenges to which has been confronted the international community for about thirty years. The fundamental challenge is as follows: how traditional knowledge, innovations and practices as regards to biodiversity can be perpetuated in a world in increasing interbreeding, and up to what point can be used and developed by the modern scientific knowledge? This work revealed that the management of the "Malambo" was basically built upon the traditional values which had allowed the local community to adapt so much to the point that some of them remained almost intact as a result of customary implementation. The respect of the tradition associated values and fear with violation of their crowned nature thus constituted the base of their protection against the anthropogenic threats and activities. Consequently, they still show regular wildlife occurrence in these days. The status of keystone species like *Cercopithecus ascanius*; *Pan troglodytes schweinfurthii,* and the *Gorilla beringei graueri* with high occurrence rates recorded within visited *Malambo* is obvious. This being said, the *Malambo* known as relics of the biodiversity in the Itombwe massif forest has remained protected thanks to the customary rules which in turn prohibited certain bad practices related to the exploitation of natural resources. The fact of creating *sui generis* legal frameworks aiming at protecting the indigenous people's knowledge is a legal obligation since the classical frameworks do not suitably help ensure this mission.

₂Corresponding author :

Keywords: *Protected area, Itombwe, Traditional knowledge, scientific knowledge, Malambo.*

Summary

The issue of protecting the traditional knowledge of indigenous peoples is at the crossroads of several major challenges that have confronted the international community over the past 30 years. The fundamental issue is: how can traditional biodiversity knowledge, innovations and practices be perpetuated in an increasingly interbreeding world, and to what extent can they be used and enhanced by modern scientific knowledge? This work revealed that the management of the *Malambo* was based on traditional values that had enabled the local community to appropriate them to the extent that some of them have remained almost intact. Respect for the associated tradition and the fear of violation of their sacredness has therefore been the basis for their protection from human threats and activities. As a result, they record high concentrations of wildlife on these days. The presence of species such as *Cercopithecus ascanius*; *Pan troglodytes schweinfurthii,* and *Gorilla beringei graueri* whose high encounter rate was recorded within the *Malambo* visited is revealing in several ways. This means that the *Malambo*, as relics of biodiversity in the Itombwe forest massif, had survived thanks to the customary rules of the area, which prohibited certain bad practices linked to the exploitation of these resources. The creation of *sui generis* legal frameworks for the protection of indigenous knowledge is a legal obligation when traditional frameworks fail to adequately address this task.

Key words: *Protected area, Itombwe, Traditional knowledge, Scientific knowledge, Malambo.*

1. Introduction

The trend towards the collapse of traditional practices linked to nature protection does not prevent us from thinking about their revival, in view of the environmental issues of the day. It is from this perspective that Roussel (2005) demonstrates the durability of local naturalist knowledge by emphasizing that "belonging to a tradition is considered a guarantee of a certain antiquity and if the elements of the biodiversity concerned have come down to us, it is because the use made of them is necessarily sustainable". Throughout human history, indigenous peoples and local communities have managed biological resources for a multitude of reasons, including subsistence, respect for nature, and cultural and spiritual purposes. This management of local resources predates modern notions of 'protected areas' by millennia and persists to the present day.

From 1980 onwards, the importance of local knowledge was taken into account in global debates on nature (Roué, 2012). This opened the way for researchers from various scientific disciplines to find a field of investigation in it (Sene, 2013). Local knowledge is increasingly called upon in interdisciplinary scientific approaches or in nature management exercises. The advent of a new type of management, starting in the 1980s, based on the interdependence between scientific knowledge, particularly ecological knowledge, and management, in the style of "knowledge for better management" (Thibault, 1991), as well as the political concept of sustainable development, has given rise to institutional and, at the same time, epistemological recognition of so-called local knowledge. One principle contained in Agenda 21 of the 1992 Rio Declaration concerns the central role given to indigenous peoples to play in the field of the environment, because of their knowledge and practices considered "traditional" (Ducros, 1998). Traditional knowledge holders are strongly connected to the natural environment. When we talk about traditional knowledge, we are also talking about biological diversity, since this is the raw material of this knowledge. So it makes no sense to talk about traditional knowledge without talking about protecting biodiversity. The 1992 Convention on Biological Diversity (CBD), adopted as a response to the rapid erosion of biodiversity, confirms this link by recognizing the value of these knowledge systems in preserving biodiversity.

Despite the rapidity of recent advances in genetics, it is essential to understand that knowledge of the properties and benefits of biological resources is not just a contemporary phenomenon. For centuries, local people around the world have acquired, used and passed on traditional knowledge about local biodiversity and its use for many purposes. From food and medicine, to clothing and construction, to the development of agricultural and animal husbandry skills and practices.

Traditional knowledge is a cumulative body of knowledge, skills, practices and representations maintained and developed by peoples whose history is interwoven with the natural environment. This sophisticated collection of understandings, interpretations and meanings is part of an even more complex cultural whole that includes language, ritual, spirituality and cosmogony.

Often referred to as traditional ecological knowledge, local knowledge, or indigenous knowledge, traditional knowledge is reflected in community practices, institutions, or in the relationship between individuals and ancestral rites. However, it is difficult to define because of its insidious and complex nature. While rooted in tradition, it also appears as knowledge in the modern sense, defined as fundamentally dynamic. Indeed, traditional knowledge is continually evolving as a response to changes in the environment.

In the context of access and benefit-sharing, traditional knowledge refers to the knowledge, innovations and practices of indigenous and local communities in relation to genetic resources. This traditional knowledge is the result of centuries of experience of people, adapted to local needs, cultures and environments, and passed down through generations.

Forest clearings are "natural" openings in the canopy of the dense forest. They are dominated by a vegetation composed mainly of herbaceous plants. They constitute particular ecosystems whose influence on the concentration of the animal population is still poorly understood. These forest clearings are wetlands rich in biodiversity, bringing together particular animal and plant associations (Vanleeuwe et al. 1998; Vande Weghe, 2006). The vegetation is semi-aquatic and the flora is largely dominated by numerous families including Cyperaceae and grasses (Boupoya, 2010). They are important centres of animal concentration for large herbivores such as elephants (*Loxodonta africana cyclotis*), buffaloes (*Syncerus caffer nanus*), gorillas (*Gorilla gorilla*), and bushpigs (*Potamochoerus porcus*), which find digestible, fast-growing vegetation available all year round (Nganongo, 2000).

This study aims to determine the role that *Malambo* play in protecting forest habitats as special ecosystems with natural openings in the canopy and open spaces as well as to highlight the relevance of forest clearings in animal population dynamics in Itombwe Nature Reserve (INR).

Studies have shown that these sites, preserved for thousands of years, are true sanctuaries of biodiversity containing plant and animal species that can be used for food, medicine, crafts, etc. It is time to gather all this local knowledge and submit it to scientific analysis in order to better integrate it into modern science. Sacred forests, often communal, are preserved as cemeteries, sanctuaries for fetishes, places of worship or initiation. The aim of this study is to demonstrate, through documentation, the effectiveness of traditional practices, beliefs and ways of life of the peoples, who have safeguarded for several centuries the natural phenomena that today represent the network of protected areas in the Democratic Republic of Congo (DRC), whose specific habitats are increasingly known thanks to science. As such, protected areas have a crucial role in maintaining a healthy environment for humans and nature. They are essential for biodiversity conservation and vital for the cultures and livelihoods of indigenous peoples and local communities (Mubalama, 2018).

Our focus in this contribution will not be on these sites in the NIR. However, with the advent of a new type of management, from the 1980s onwards, based on the interdependence between scientific knowledge, particularly ecological knowledge, and management, of the "know to better manage" style, it is essential to have an adapted representation of innovation in order to better understand how the joint development of innovations takes place (Thibault, 1991). It is not our intention to explain the interest that these sites represent for a history of landscapes that has already been well explored, but rather to raise questions that arise today more than ever in terms of their conservation, based on our recent fieldwork and elements of understanding of the issues to which these sites are subject today, taking into account the analysis of the historical context linked to the inclusion of.

With this in mind, this paper aims to address the concerns raised above with particular emphasis on the following points: (i) to take stock of ongoing research on cultural practices for safeguarding and conserving plant diversity in the Itombwe Forest and the threats, challenges and opportunities; (ii) to develop a concerted methodology for research, exchange and dissemination of good practices; (iii) to propose strategies for embedding local and national biodiversity use policies, plans and programmes in local and national cultures; and finally (iv) to devise new avenues of research.

2. Materials and Methods

2.1. *Location of the study area*

Located in the Province of South Kivu in DRC, the NIR is an IUCN category VI protected area located in the Itombwe forest massif (Figure 2) with an area of 573,222 ha (Mubalama et al. 2013; Mubalama et al. 2017). A peculiarity observed in the distribution of animal species within this protected area raised the curiosity to characterize the ecological and socio-cultural potentialities of their habitats. From an ecological point of view, the *Malambo are* perceived by the local population as "maternity wards" (Hart & Mubalama, 2005) because some females migrate there at the time of calving to gain easy access to rock salt. The NRT was created by Ministerial Order No. 038/CAB/MIN/ ECNEF/2006 of 11 October 2006 and completed by Provincial Order No. 16/026/GP/SK of 20 June 2016 on provisional measures to update the boundaries resulting from the participatory delimitation of the NRT. At this stage, 41.5 km of the external boundaries of the NRT have been demarcated using signs including 12 km on land based on geospatial boundaries or confluences, straight lines or parallels and 29.5 km of natural boundaries following the hydrographic network including the Mwana, Kikuzi, Mazimuilu and Mulambozi rivers in Basile and Wamuzimu chiefdoms. (WWF, 2018).

The Itombwe Massif is the largest group of sub-montane forests in Africa (Doumenge, 1998). The Itombwe Massif covers an area of about 12,000 km2 and includes a single forest zone, the Central Massif, covering 6,500 km2 with an altitude of up to 3,475 m at Mount Muhi.

The Central Massif contains several unique forest habitats: bamboo, heathland, wet transitional zones and a high altitude ecotone tree savannah. There are significant populations of Eastern Chimpanzees (*Pan troglodytes schweinfurthii*) and the threatened Grauer Gorilla species (*Gorilla beringei graueri,* Critically Endangered Species. http://www.iucnredlist.Org/details/39995/0, accessed 11/11/2016).

The Itombwe Massif represents one of the most suitable areas for the conservation of great apes and the endemic biodiversity of the Albertine Rift Massifs. The Itombwe Massif is also one of the priority sites for its biological importance at the global level and at the level of the Albertine Rift (Doumenge, 1998).

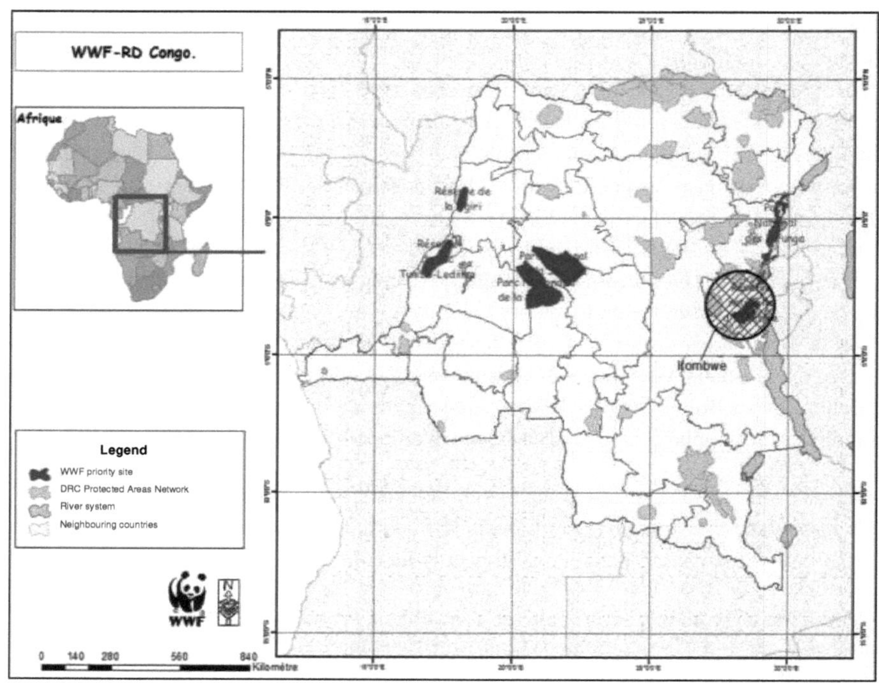

Figure 1. Location of Itombwe Nature Reserve

2.2. Presentation of forest clearings

Clearings play a role in maintaining biological diversity. They vary in size from a few square metres to hundreds of hectares. They are found on a variety of soil substrates ranging from deep, marshy soils to thin soils on rocky outcrops. Letouzey (1985) identifies them as grasslands developed on chlorite schists of the poorly metamorphosed Precambrian series, the study of which is still very incomplete. The Itombwe Mountains are an integral part of the mountainous chain stretching across the eastern border of Congo, from Katanga to North Kivu. This chain borders the tectonic gap of the western branch of the Great African Rift, the bottom of which is occupied by several large lakes (Jeune Afrique, 1978, 1993). Two major geomorphological complexes can be distinguished: the plateaus of the eastern half, dominated by the mountains of the northern and eastern margins, and the slopes and fluvial plains of the western half. The complexity of the soils of the Itombwe Massif makes it very difficult to describe in a brief note. This complexity is due to the variety of the original rocks, the relief and the climates. The undulating highlands derived from Burundian rocks are covered with acidic soils of low fertility that can be transformed into lateritic cuirasses; soils in the valleys and foothills are more fertile (Wilson, 1990; Doumenge, 1998; Muhigwa et al. 2011). These clearings are characterized by predominantly herbaceous vegetation in which the grass family is both the most abundant and dominant, giving them their role as natural pasture. The species Myrianthus *arboreus* P. Beauv, *Symphonia globulifera* (Lin) f, *Macaranga mildbraediana* Pax, *Harungana Montana* Spirlet and *Parinari excels* Sabine are the most dominant and abundant (Table 2) and thus influence the physiognomy of the clearings. The determinism of the formation and evolution of clearings is poorly understood. Aubréville (1948) assumed that the grassy formations of forest clearings are of recent origin and are obviously the result of human activity only. He also postulated that forest clearings or grassy savannahs in forest environments would represent paleo-climatic relics. Letouzey (1985) states that no ancient human action is evident around these formations and their anthropic origin seems not to be retained.

Forest clearings are still influenced by the local geology and the current climatic and soil conditions. They play diverse functions for biological communities and constitute ecosystems whose ecological role has not been fully explored to date.

Several animal species, including those listed in the IUCN appendices and others fully protected in DRC, are only found around multiple swampy clearings that the local population of the area calls *Malambo* (Figure 2 & Table 1). The survey results revealed that *Malambo* are nothing more than swampy clearings or wetlands. Seven of these were visited during the course of this work. They have been protected by local tradition and custom since ancient times. Having for this purpose the sacred character, they are almost totally excluded from the hunting areas.

Figure 2. A view of Ilambo M'banga not far from the Kikuzi River overlooked by a rainforest edge in Groupement Bashimwenda [1er], Mwenga.

Because of the mysteries that surround them and the deep fear that they arouse in people, these places are peaceful. They represent privileged environments for spawning, refuge and nursery for certain aquatic species of the zone. This has allowed us to observe, until today, strong concentrations of animal species. Thus, the presence of certain saline rocks observed in most of them, is the basis of their designation as salines (Ibucwa, 2008).

Forest clearings are the result of the influence of several complex and dynamic factors over time. Their origins are varied and the determinism of their formation is complex. They are dependent on various factors and we can mention for example: (i) the predominant geological formations; (ii) the pedological formations on which they have developed; (iii) the elements of the local climate; (iv) the dominant phytosociological composition; (v) human activity, notably with the bushfire element which has an impact on the vegetation of the clearing and which can also be broken down into several factors (period of burning, duration of burning, type and state of the burnt vegetation, ...). They remain, as grassy savannahs in forest environments, influenced by local geology and current climatic and soil conditions (Noupa & Nkongmeneck, 2008).

2.3. Methodology of the work on the biological knowledge of the Malambo

2.3.1. Plant identification and sampling
Twenty-two clearings on hydromorphic soil were identified in the field. Seven of them were selected for this study. This choice was made on the basis of distance from the RNI headquarters in Mwenga, accessibility and the logistical means at our disposal. The size of the clearings, degree of water saturation, shrub cover, and the presence or absence of animal activity were also taken into account in this choice.

The identification of species was an important step in this work. A first identification was made in the field from forest botany manuals, practical works on the identification of the flora of Central African species (Letouzey, 1982, 1983).

Pocket herbariums were also made *in situ* for comparison with previously collected and identified specimens, as it is impossible to take the rare flora available in the field with us. We also benefited from the experience of local botanists in the field. The determinations were made on the basis of existing floras: Flora of Congo, Central Africa, and Flora of East Tropical Africa.

The search for documentation and bibliographical references were part of the essential basis in the production of this article. The documents used here are mainly based on the literature, surveys and polls as well as the analysis of soil science samples from the Faculty of Agronomy of the Catholic University of Bukavu (UCB), the *herbarium* of the Centre for Research in Natural Sciences (CRSN/Lwiro) and the consultation of some websites deemed reliable. The online Ramsar fact sheet, the guidance documents on rapid cultural inventories of wetlands and the designation sheet for wetlands of international importance were used to develop the survey questionnaire.

The survey was conducted among the local community in the form of a rapid cultural inventory as required by the Ramsar Convention (Pritchard, 2010). It aims to collect relevant information on the perception of the local communities on the wetland theme. The survey included two categories of questions: open-ended questions, where the interviewee had to answer with his or her own ideas, and a second, closed-ended question, where the interviewee had to choose an answer from among the proposed assertions. This information was collected for the period from 04 to 24 July 2016 (Igunzi, 2017).

2.3.2. *The phytosociological method of Braun-Blanquet*
This method is based on the principle of discontinuity, where the vegetation is formed of well-defined and homogeneous floristic associations. Their characteristics and distribution are based on ecological and phytocoenotic factors. This fundamental concept, developed by the Zurich-Montpellier School (Braun-Blanquet, 1932), considers the association as a homogeneous floristic grouping defined in an abstract way on the basis of the synthetic comparison of floristic lists established on structurally and floristically homogeneous surfaces. The notion of characteristic species, i.e. species exclusively present in an association, remains fundamental because it serves to define the different associations. This method has been used by many authors for the first studies of tropical vegetation, such as Lebrun (1947), Duvignaud (1949), Léonard (1950, 1952), & Troupin (1966).

Although the application of this method in tropical forests has been the subject of much criticism (Aubreville, 1951; Devred, 1961), the azonal character of our study sites allowed its application because the vegetation is composed mainly of specialized herbaceous species. Therefore, the Braun-Blanquet method is applicable for the characterization of intra-forest clearing groups on hydromorphic soil composed essentially of Cyperaceae and Poaceae to which are added various other species.

Seven *Malambo were* visited (Figure 2); the list was later supplemented with SMART (*Spatial Monitoring and Reporting Tool)* data from NRT eco-guard patrols. With regard to their ecology, the study focused on seven that were subject to botanical, pedological and hydrological sampling. These seven *Malambo* (Table 1) constitute what we call in this work the "study site", abbreviated to "ES". Each was given a serial number (e.g. SE-01); a brief phytosociological study was also carried out and geo-referenced coordinates taken using Garmin 64s GPS (Table 1).

Table 1: Geo-referenced coordinates of sampled Malambo

N°	Local name of *Ilambo*	Location coordinates in UTM				Distance to - next (Km)	Code	Nearby landmark
		Longitude	Latitude	Altitude (m)	Average depth			
01	ALELE	0695874	9639810	2183	48 cm	0	SE-01	Riv. Asangye
02	ATOBO	0671857	9640036	1909	28 cm	24,02	SE-02	Riv. Kikuzi
03	ESSESA	0671547	9640056	1902	41 cm	0,31	SE-03	Riv. Kikuzi
04	IBU	0672288	9640374	1959	25 cm	0,81	SE-04	Riv. Ibu
05	KALAMBO	0681120	9646450	1890	30 cm	10,72	SE-05	Anc. Road
06	LUSASA	0694268	9632690	1945	17 cm	19,03	SE-06	Riv. Lwelaki
07	M'BANGA	0671445	9640363	1896	19 cm	24,08	SE-07	Riv. Kikuzi
	Total					**78,97**		

The study was conducted within the clearings, taking into account the homogeneity of the vegetation. The average depth was estimated by pushing in different places the graduated bar. The inventory of woody species was carried out in circular plots of 2.5 m radius each where the counting was done following a linear transect of 1.5 km. The inventory mainly concerned woody species with a superficial look at herbaceous species. To quantify the representation of woody species suggests measuring their diameter at the height of the human chest called DBH. Thus, a total area of 137.37m2 was covered for the seven clearings. Three parameters were studied for each species: basal area (St), relative density (DER) and relative dominance (DOR).

Relative density (DER): *This took into* account the number of individuals observed within each *Ilambo*. Where, *Ni* = the number of individuals of a species or family, and *Y,Ni I*= The total number of individuals in the sample.

$$DER = \frac{100 \times N_i}{\sum_i N_i}$$

Basal area (St): It took into account the occupation of a species or family in each *Ilambo*. With, D = Diameter of the individual considered at the level of the DBH, and is expressed in m2/ha.

$$S_t = \frac{\pi \times D^2}{4}$$

Relative dominance (DOR): It highlights the species or families occupying more space within the study sites. Where, St = basal area of a species or family, and SSti *I* = total St of individuals in the sample.

$$DOR = \frac{100 \times S_{t_i}}{\sum_i S_{t_i}}$$

The determination of known plant species was done on site. And for those to be confused, it was done with the help of the work of (Lebrun, 1991), or by identification at the *herbarium* of the Natural Sciences Research Centre of Lwiro. Finally, everything was updated either from the *world check-list* of the Royal Botanic Gardens of Kew, or from the *Conservatoire et jardin botaniques de la ville de Genève* website.

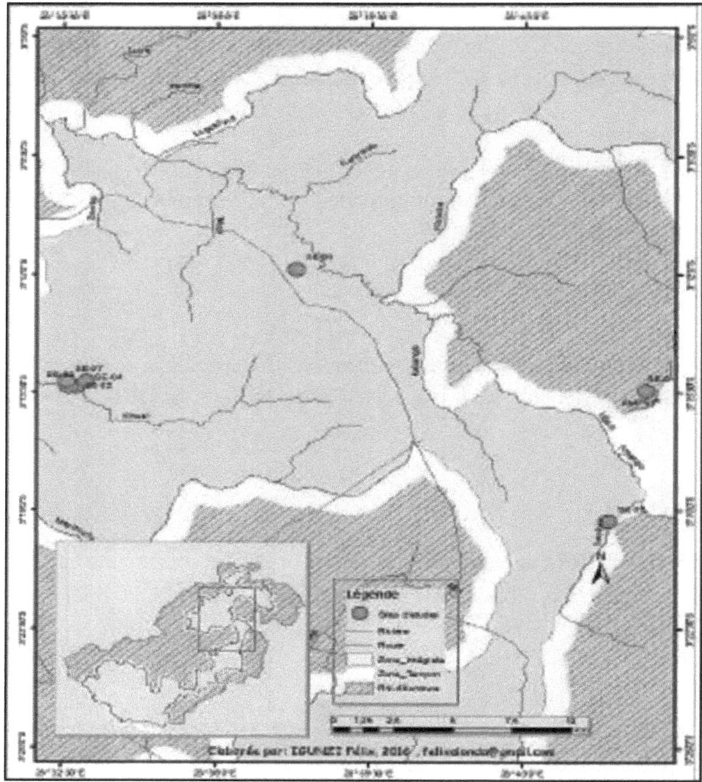

Figure 2: Map showing the study sites in Itombwe Nature Reserve

Three soil samples were taken at each *Ilambo* depending on whether the location (rock) showed evidence of being more licked by certain species. After mixing the three samples, a single sample of at least 0.5kg was maintained. This was transported in a sterile bag to the soil science laboratory of the Catholic University of Bukavu, where 17 physico-chemical parameters were analyzed.

It should be noted that other recent wildlife clues were recorded along the route separating the different wetlands visited, associating a geographic coordinate with each clue. These clues included species droppings, food crumbs, tracks, calls heard or recent nests observed. They were analysed and/or interpreted from several angles, including the rate of encounters as a function of the distance travelled.

The objective of variable classification is to construct classes of variables that are strongly related to each other and thus remove redundant information. The approach *ClustOfVar* used provides simultaneously groups of variables as well as the synthetic variables associated with the classes of variables. In this algorithm, the homogeneity criterion is based on the notion of correlation for quantitative variables and correlation ratio for qualitative variables. The variable classification step allows us to obtain synthetic variables that we propose to read as a kind of gradient. In our data, the values correspond to distinct groupings of modalities that are relevant for the interpretation. This approach allows us to read and label each synthetic variable. In this way, we highlight the trends that will determine the farmers' opinions regarding their consideration of the environment. Then, we specify these results by carrying out a classification on the scores of the individuals measured on the synthetic variables. From a sociological point of view, the contribution of the synthetic variables to the interpretation of the standard profiles obtained is obvious.

2.3.4. Environmental parameters

The degree of hygromorphy of the soil was estimated according to the scale proposed by Senterre (2005): 1 = low, well draining soil, quickly rehydrating after the rains; 2 = medium, soil does not rehydrate quickly after the rains; 3 = high, periodically flooded soil remaining saturated with water for a long time after the rains; 4 = permanent, soil saturated with water even outside the rainy season.
The intensity of animal frequentation is obtained by the encounter rate (ER) of animal tracks (prints, droppings, tracks).

Phytogeographic types were identified for each species with reference to the phytogeographic subdivisions of Africa by White (1983). The distributions of the species were determined from the world check list of the Royal Botanic Gardens, Kew (http://apps.kew.org/wcsp/home.do) and from the distributions given in the flora and articles cited above for species identification. The biological types of each species were defined according to the model of Raunkiaer (1934).

2.3.5. Different approaches to local knowledge

There are three main approaches to interpreting local knowledge: (i) *local knowledge as a legacy of the past:* This approach shows the kind of reverence one should have for the accumulated wisdom of past generations, so intensely expressed in Amadou Hampaté Bâ's famous phrase, "every time an old man dies, a library burns"; (ii) *local knowledge as an embodiment of a different and specifically African way of thinking:* this is an African "epistemology" and thus a means of rethinking development methods in areas such as health, agriculture and natural resource management.

Proponents of this approach point to the failure of current development methodologies as evidence of the need for new concepts rooted in the cultural heritage of the populations concerned (Bérard et al. 2005; Tubiana, 2005) and finally, (iii) *local knowledge as a means and process of expressing what indigenous populations know, and a means of involving them in the acquisition of the knowledge required for development*, and thus passing on to future generations the best the present has to offer. Proponents of this approach insist that promoting local knowledge is as much a matter of empowering local actors to produce new knowledge based on both the heritage of the past and a lucid assessment of current challenges, as it is of simply cataloguing and storing the traditions inherited from the past.

3. Results and discussion

3.3. *Ecological potential of the Malambo in the NIR*
The *Malambo* is home to the IUCN Red List species *Terathopius ecaudatus*, and also hosts *Phataginus tricuspis*, a near-threatened species. Other rare animal or plant species and threatened ecological communities (i.e., endangered, critically endangered and vulnerable species) justifying the international importance of *Malambo* as a wetland in the Itombwe Forest Massif are detailed in Table 3 respectively.

3.4. *Characteristics of the attraction of clearings on wildlife species*
This analysis demonstrated the role of clearings in wildlife population aggregation. The clearings are points of concentration for the wildlife population. From the data collection sheets, it was found that the tracks of 16 species were found in the inventories made in and around the clearings of Alele, Atobo, Essesa, Ibu, Kalambo, Lusasa and M'banga (Figure 3 &Table 1). The calculation of specific densities for these different species according to the type of environment (clearing, swamp forest or heterogeneous forest) was carried out in order to quantify the representation of woody species (Figure 4 and Table 2). The results show that the clearings visited are characterized by predominantly herbaceous vegetation in which the Cyperaceae and Grass families are both the most abundant and dominant (Figure 4), which gives them their role as natural pastures. As such, they are important centres of animal concentration for large herbivores, which find there digestible vegetation that grows rapidly and is rich in mineral salts (sodium, calcium and magnesium) and water, available throughout the year.

The peak of the genus *Cephalophus spp.* can be distinguished, which represents the cumulative density of 4 species of Cephalophus: *C. ascanius, C. mitis, C. lhoesti, C. denti.* These are fearful species, which are easy prey for predators that frequent areas of high animal concentration. For this reason, duikers preferentially evolve in heterogeneous forests around clearings, making only sporadic incursions to graze in forest clearings. As also documented in Boumba-Bek National Park (southeast Cameroon), the species *Gorilla gorilla gorilla* (gorilla), and *Potamochoerus porcus* (bushpig), also show high densities in heterogeneous forests. These are in fact ubiquitous species, which prefer all habitats in the environment according to their food needs. This is not the case for *Syncerus caffer nanus* (buffalo) and *Tragelaphus euryceros* (bongo), which are species that are strongly attached to forest clearings where they spend most of their time, moving only in search of water when there is none in the clearings.

The densities of these species are clearly higher in the clearings than those of other species. This shows that the density of tracks calculated in the different inventory segments in the 7 clearings follows a gradient that decreases as one moves away from the centre of the clearing. This indicates that the animals (all species) are not uniformly distributed in the forest massif around the forest clearing. This reflects the influence of the forest clearing on the distribution of the animal population in the adjacent forest, as forest clearings are rich in plant species that are highly valued by herbivores. In addition, the richness of the clearing in mineral salts (sodium, calcium and magnesium) and water constitutes a source of food supplement for these animals. Moreover, the minimum production of fruit in the dry season constitutes a real shortage and the survival of the vertebrate community (herbivores and frugivores) depends on the fruiting of these trees. However, in forest clearings, food resources are in large numbers.

Animal visitation rates showed that two species in particular, *Cercopithecus ascanius, Chimpanzee* and *Gorilla,* seem to find all the conditions favourable to their presence in these clearings. Indeed, the frequency rates of these three species were the highest of the 9 species identified. On the other hand, the other species presented frequency rates lower than 5% (Table 3). The high presence of chimpanzees in this clearing could be explained by the fact that these animals are widely distributed in the Reserve (Mubalama et al. 2017) and are attached to these environments where they spend most of their time. These results hardly corroborate those obtained by Noupa et al. (2008) who showed that it is rather the species of the genus *Tragelaphus* that display a positive tropism for forest clearings and wetlands. Several studies have shown the major role that clearings play in the ethology of different animal species that find in these environments the elements necessary for their survival (fast growing digestive vegetation, resting place, grazing place, etc.). (Nganongo et al. 2000; Noupa, 2008).

The protection of the *Malambo is the* result of actions arising from the prohibitions dictated by traditional custom, which are very restrictive but have the advantage of allowing the guardians of sacred places to retain influence over the management of sacred spaces, including the *Malambo*. Given the importance of sacred forests, the DRC has taken them into account in the 2002 Forestry Code Act. Forests belonging to local populations are recognised as having a vocation for conservation and protection (Luketa, 2005).

Customary laws govern the traditional management system that can enrich national biodiversity policy. Thus, the CBD, which came into force in 1993, recognizes the "sovereign right" of States to manage their genetic resources and to preserve their traditional knowledge, which has been elevated to the status of "common heritage of humanity" (Larrère, 1997). Indeed, during the 4th session of the conference on the ecosystems of dense and humid forests of Central Africa, held in Kinshasa (DRC) from 10 to 13 June 2002, it was recommended that sacred forests be taken into account in any forest management process.

As is well known, local knowledge is now at the heart of the issue of rational management of natural resources. While its recognition is certain within the international community, its practice does not seem to be easy. Taking indigenous knowledge into account in the sustainable development approach remains the terrain of many amalgams and conflicts of interest. The challenge facing the social sciences is to come to grips with this knowledge and to find paradigms that can provide stakeholders with the tools to create a symbiosis in the intervention of the various actors. This is an opportunity to be seized by the social sciences in order to assert themselves further in the development process. In order to facilitate the integration of traditional management systems into modern contexts, the State should take stock of traditional knowledge and strengthen collaboration between traditional authorities and State institutions. Non-governmental organizations that focus on heritage conservation need to support the efforts of those working to protect ancestral values. This can take the form of multi-faceted support such as the provision of human resources (trainers) and logistics to enable them to combat all forms of smuggling.

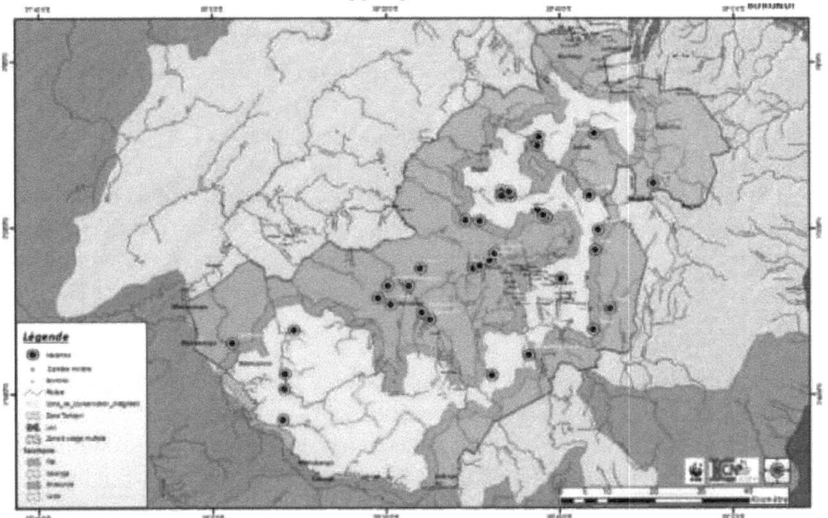

Figure 3: Spatial distribution of specific clearing habitats known as Malambo in Itombwe Nature Reserve, Source: ICCN-WWF, 2018)

These clearings play different roles in the ethology of the different species that frequent them: nuptial arena, grazing site, resting place, etc. In view of the interdependencies in the feeding relationships between the different animal species in the area, forest clearings are the only grazing sites for herbivores, and they concentrate around them all the complexity of the trophic chain in the forest environment. Their role as a focal point for animal concentration is thus highlighted by the present study.

There is an urgent need for a study on the dynamics of these particular ecosystems, which are the basis for the stability of animal populations in the Congo Basin forest massif, with a view to improving their value for wildlife management. Despite the short duration of the study, which could be considered a limiting factor in the observation of the number of animals, the daily activity of the animals, in particular those of *cercopithecus ascanius* and the great apes, could be observed. For both species, the day's weather had no influence on their presence or absence in the clearings (the presence rate for *cercopithecus ascanius* was 61.61% compared to 14.29% for chimpanzees and 7.14% for eastern lowland gorillas).

As much as wetlands are the object of specific and particular interventions, it is urgent that a "forest clearings in the Congo Basin" project be developed in order to develop a coherent strategy for the management of these ecosystems, which are animal concentration points and therefore vital centres in the same way as the water points in the massif, given that no evidence denies their role in the determination of large migratory species. Do forest clearings play the role of relays in biological exchanges? If so, special provisions should be made in the master plans for the development of the forests for the clearings located in the protected areas as ecosystems of vital sites indispensable for their survival.

3.5. *Physico-chemical, soil and hydrological parameters and implications for NRT conservation*

Following a physico-chemical analysis carried out on samples of water and soil in the *Malambo*, the results revealed that the appreciation of these biotopes by several animal species is not only linked to the quietude established by custom and tradition, but also to the high salt content of the soil and water in some of them, factors that attract several species to drink in specific places. Although the vegetation is rich in woody and herbaceous plants, the latter has not revealed any particular element of attraction that would justify the abundance of wildlife.

Despite these significant faunal and floristic potentialities, the NRT has no international status to date. A comparison of the ecological and socio-cultural potential of the *Malambo with* the criteria of the Ramsar Convention showed that these marshy clearings, as true wetlands, meet up to 44.4% of the criteria for inclusion in the Ramsar list. This strategy when implemented will allow the elevation of the NRT as a wetland of international importance and will result in the conservation of its biodiversity in the long term. From a socio-cultural point of view, these areas have a sacred character which makes them, in majority, no hunting areas (Defailly, 2010).

Table 2: Woody species justifying the floristic importance of Malambo

Species	Family	NI	ST	DOR	DER	DBH
Myrianthus arboreus P. Beauv.	Moraceae	21	0,02	1,07	17,07	0,15
Symphonia globulifera (Lin) f.	Clusiaceae	18	0,03	1,72	14,63	0,19
Macaranga mildbraediana Pax	Euphorbiaceae	15	0,02	1,38	12,20	0,17
Harungana Montana Spirlet.	Hypericaceae	14	0,08	4,89	11,38	0,32
Parinari excels Sabine	Chrysobalanaceae	10	0,13	7,63	8,13	0,40
Entandophragma excelsum Spague.	Meliaceae	8	0,15	9,24	6,50	0,44
Galiniera coffeiodes Delile	Rubiaceae	7	0,30	18,34	5,69	0,62
Strombosia scheffleri Engl.	Olacaceae	5	0,09	5,51	4,07	0,34
Chrisophylum albidum G. Don	Sapotaceae	4	0,24	14,43	3,25	0,55
Grewia malacocarpa Mast.	Malvaceae	4	0,03	2,10	3,25	0,21
Ocotea usambarensisEngl.	Lauraceae	4	0,09	5,20	3,25	0,33
Xymalos monospora(Harv) Bail	Monimiaceae	4	0,13	7,63	3,25	0,40
Polyscias fulva (Hiern). Harms	Araliaceae	3	0,13	8,02	2,44	0,41
Sapium ellipticum (Hochst) Pax	Euphorbiaceae	2	0,07	4,01	1,63	0,29
Bridelia micranta (Hochst) Bail	Euphorbiaceae	1	0,02	1,07	0,81	0,15
Ficalhoa laurifolia Hiern.	Sladeniaceae	1	0,01	0,48	0,81	0,10
Syzygiumcongolense Vermoesen.	Myrtaceae	1	0,07	4,29	0,81	0,30
Trema orientalis(Lin) Blume	Cannabaceae	1	0,05	2,98	0,81	0,25
Total		123	1,65	100,00	100,00	5,62

The wetlands inventoried justify this criterion by the results in Table 2, which reflect their floristic richness. It follows that apart from the Cyperaceae and Poaceae species dominated by *Imperata cylindrica* and *Pennisetum* sp. and the undergrowth, the wetlands in the Itombwe forest massif contain up to 16 families of woody species grouped into 18 genera.

In particular, the species *Myrianthus arboreus* presents a high relative density in number of individuals observed. The families with high floristic density can be classified as follows: *Moraceae* (17.07%), *Clusiaceae* (14.63%), *Euphorbiaceae* (12.20%), *Hypericaceae* (11.38%) and *Chrysobalanaceae* (8.13%). It follows from this result that *Malambo* presents a floristic richness. The confrontation between woody and herbaceous species recorded within the *Malambo* in Itombwe forest massif, reflects on their degree of similarity (Igunzi, 2017).

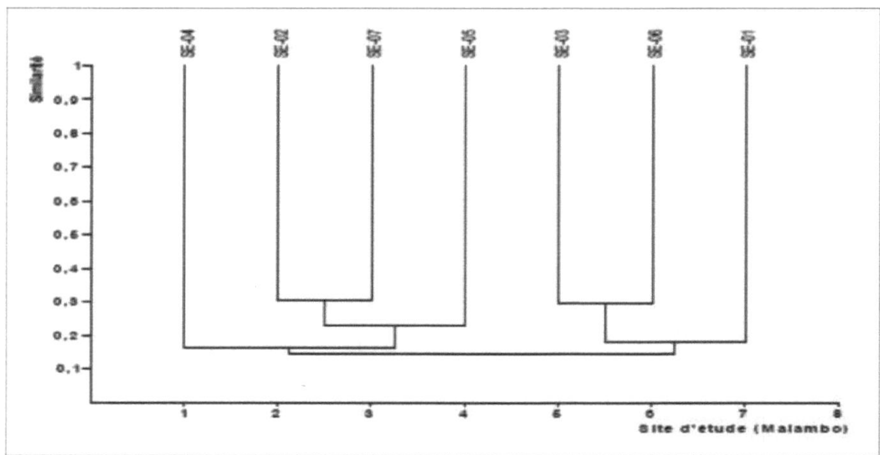

Figure 4: Degree of floristic similarity between the seven Malambo visited.
As a result, the flora that characterizes the wetlands in the Itombwe Forest is grouped into two distinct blocks. The proportion of similarity varies approximately between 15 and 30%. This low similarity, that is, high floristic dissimilarity is explained by the dominance of herbaceous species of the Cyperaceae family (*Cyperus* sp) that characterize some *Malambo* (Igunzi, 2017). Others being essentially dominated by woody species. Only one wetland (SE-05) showed mixed vegetation.

Table 3: Wildlife encounter rates at study sites

N°	Species (scientific name)	Non vernacular	Types of observation						Total obs.	Txr/ Km	%	IUCN status
			Viewed at	Poop	Cries	Trace	Nest	Miett e.				
1	*Cercopithecus ascanius*	Cercopithecus	19	20	9	4		17	69	0,20	61,61	VU
2	*Gorilla beringei ssp graueri* Matschie, 1914	Gorilla (*)	-	1	-	-	7	-	8	0,47	7,14	CR
3	*Hystrix africaeaustralis* Peters. 1852	Hog- spy	-	-	-	5	-	-	5	0,09	4,46	LC
4	*Neotragus batesi.* Winton 1903.	Antelope	-	3	-	-	-	-	3	0,14	2,68	LC
5	*Pantroglodytes. SchweinfurthiiGiglioli,* 1872	Chimpanzee (*)	-	2	-	4	10	-	16	0,21	14,29	IN
6	*Potamochoerus porcus.* Linnaeus, 1758.	Bushpig	-	-	-	1	-	-	1	0,12	0,89	LC
7	*Psittacus erithacus* Linnaeus. 1758	Parrot (*)	4	-	-	-	-	-	4	0,52	3,57	VU
8	*Tryonomis sp*	Gambia Rat	2	-	-	-	-	-	2	0,17	1,79	-
9	*Bitis gabonica* Duméril 1854.	Viper	4	-	-	-	-	-	4	0,06	3,57	-
	Total		29	26	9	14	17	17	112	-	100,0 0	

The importance of *Malambo* for wildlife species was highlighted by considering physico-chemical parameters of water and soil samples. The aim is to understand the reasons justifying the particular appreciation of this water and/or soil by certain species in the Itombwe forest massif. Indeed, the clearings of Itombwe forest massif are characterized by a predominantly herbaceous vegetation in which the grass family is both the most abundant and the most dominant, which gives them their role of natural grazing (Igunzi, 2017). The results in Table 3 highlight the intensity of animal use observed in different wetlands in Itombwe forest massif.As a result, 112 observations were recorded along the route, based on direct observations (sightings), or indirect observations (recent droppings, calls, recent tracks, nests or food crumbs). Nine animal species were recorded during this survey as being more frequent in the *Malambo*. In the top three positions were Cercopithecus (61.61% of the encounter rate), followed by Chimpanzee (14.29%) and Gorilla with (7.14%), etc.

Table 4: Results of analysis of soil elements sampled in the Malambo

N°	Site code.	Laboratory number.		pH-Kcl	Conductivity electrical(qS/c m)	Organic Carbon (%)	Total nitrogen (%)	Assimilable phosphorus	CEC (cmol/Kg)	H+Al (cmol/Kg)	Al3+ (cmol/Kg)	H+ (Cmol/Kg)	Potassium (cmol/Kg)	Sodium (cmol/Kg)	Calcium (cmol/Kg)	Magnesium (cmol/Kg)	Clay	% Sand	Silt
01	SE-01	ZO 3	5,65	5,32	449	24,66	2,09	22,28	32,39	0,036	0,32	-	1,51	2,97	6,96	3,84	36,2	47,8	16
02	SE-02	K3 3	4,67	3,73	80,7	2,68	0,046	37,73	10,59	4,7	1,12	3,58	4,4	12	19,59	2,99	12	80	8
03	SE-03	K3 2	5,23	4,70	98,3	0,72	0,11	31,32	3,84	0,1	0,40	-	2,4	9,2	15,75	3,83	6	92	2
04	SE-04	K3 4	5,49	4,81	34	0,59	0,23	27,84	3,36	0,5	0,56	-	3,4	8,8	14,69	3,58	4	76	20
05	SE-05	ZO 1	3,72	3,56	137	5,42	0,44	28,57	10,41	2,66	1,68	0,98	0,31	0,87	1,53	2,09	8,2	81,8	10
06	SE-06	ZO 2	5,77	5,33	195	3,64	0,30	23,2	8,82	0,072	0,24	-	1,67	3,89	5,36	2,64	10,2	73,8	16
07	SE-07	K3 1	4,93	4,54	167	1,45	0,30	54,74	6,36	0,5	0,48	0,02	2,8	11,8	16,63	5,10	6	64	30
	Average		**5,07**	**4,57**	**165,86**	**5,59**	**0,50**	**32,24**	**10,82**	**1,22**	**0,69**	**1,53**	**2,36**	**7,08**	**11,50**	**3,44**	**11,80**	**73,63**	**14,57**

Table 4 first presents the results for each parameter analyzed at each wetland individually.) Table 5 also presents, individually, the results of each water parameter analyzed at each wetland. Then, it presents the average value of the same parameter for the three wetlands whose samples were analyzed. These results therefore provide information on the overall physicochemical composition of the water in the wetlands visited in Itombwe Forest.

The results of the soil analysis of the *Malambo* presented provide information on the overall physico-chemical composition of the soil in the wetlands visited in the Itombwe forest massif. The following can be deduced from this: (i) the pH varies from 3.72 to 5.77 with an average of 5.1, i.e. an acid pH, as it is lower than 6; (ii) the pH-Kcl varies from 3.73 to 5.33, with an average of 4.57; (iii) the organic carbon varies from 0.59 to 24.66% per site with an average of 5.59, which shows a high content of organic matter (Content > 3%). High organic carbon levels imply limitations in soil organic matter; (iv) total nitrogen ranged from 0.046 to 2.09% per site with an average of 0.5 indicating high nitrogen content (> 0.25%); (v) available phosphorus ranged from 22.28 to 54.74 ppm with an estimated average of 32.24 ppm (> 25 ppm) confirming a high level of phosphorus in the soil. 25 ppm) confirming a high phosphorus content; (vi) potassium (cmol/Kg) varies from 0.31 to 4.4 cmol/kg with an average of 2.36, indicating a high potassium content in six wetlands visited (SE-01, SE-02, SE-03, SE-04, SE-06 and SE-07) with the exception of one (SE-05) with a low potassium content (0.31 < 0.45); (vii) Sodium (cmol/Kg) varies from 0.87 to 12 cmol/kg with an average of 7.08cmol/Kg showing a high sodium content (Content >2 cmol/Kg) with a taste of water considered unpleasant because the concentration of sodium exceeds 200 mg/l (Beed & Myers, 2000); (viii) calcium (cmol/Kg) ranging from 1.53 to 19.59 cmol/kg, with an average of 11.50 showing high calcium content (Content > 5) except for the. Wetland (SE-05) which showed low calcium content (Content < 5) and finally, (ix) magnesium (cmol/Kg) varying between 2.09 and 5.10 cmol/kg, with an average of 3.44cmol/Kg, i.e. high magnesium content (Content >1.5 cmol/kg).
Also, overall, the cation exchange capacity (CEC) varies from 3.36 to 32.39 cmol/kg, with an average of 10.82. These results show that the soils of the four *Malambo*, namely: SE-03, SE-04, SE-06 and SE-07 are light (Contents <15). They have a low CEC, which can be explained in part by the sandy texture of the soil that characterizes the environments visited, while the soils of SE-01, SE-02 and SE-05 had a fairly high CEC, i.e., average soils (>15%).
The degree of hygrometry for seven *Malambo* corresponds to scale 4, according to the hygrometric classification proposed by (Senterre, 2005). It follows that the wetlands in the Itombwe forest massif have permanent soil, i.e. saturated with water even outside the rainy season.

Table 5: Results of analyses of hydrological elements sampled in the Malambo

N°	Code du site	N° du labo.	pH-H₂0	Conductivité électrique (qs/cm)	Turbidité(UNT)	Dureté totale (g/l)	Phosphore total (ppm)	Chlorure (mg/l)	Sulfate (mg/l)	Fluorure (mg/l)	Nitrate (ppm)	Ammonium (ppm)	Potassium (mg/l)	Sodium (mg/l)	Calcium (mg/l)	Magnésium (mg/l)	Bicarbonate (mg/l)
01	SE-01	Z03	6,51	432	129	376,69	0,052	532,5	64,464	3,40	31,2	0,22	169,26	4986,4	1048	265,2	520
02	SE-05	Z01	4,93	122	27,9	32,196	0,012	532,5	51,072	2,51	27,4	0,12	4,29	280,6	250	49,2	550
03	SE-06	Z02	5,84	100	727	51,51	0,041	710	95,568	2,34	30,5	0,16	5,85	230,0	1024	189,6	420
	Moyenne		5,76	218,00	294,6	153,47	0,04	591,67	70,37	2,75	29,70	0,17	59,80	1832,3	774	168,0	496,67

3.6. Comparison of some physico-chemical parameters of water and soil samples in *Malambo*

The results in Table 6 reflect a clear comparison between different concentrations in the *Malambo* water and soil. The average contents obtained for the seven different physico-chemical parameters in the water and soil show that the *Malambo* water is more concentrated than the soil. Indeed, comparing the contents of (pH), potassium (K), sodium (Na), calcium (Ca) and magnesium (Mg), we notice that the values of water prevail over those of the soil.

Table 6: *Comparison between physico-chemical parameters of water and soil in Malambo*

N°	Parameters Samples (average)	analysed	pH	Conductivity Electrical.	P (mg/l)	K (mg/l)	Na (mg/l)	Ca (mg/l)	Mg(mg/l)
01	Water	3	5,76	2180	,04	59,80	1832,3	774	168
02	Soil	7	5,07	165,86	32,24	2.36	7,08	11,50	3,44

3.4.1. *Other parameters of the Malambo water compared to the drinking water*

Since the water in these clearings is more or less loaded with mineral salts (Magliocca & Gauthier-Hion, 2001), the "saline" effect is an additional attraction for this fauna. Vande weghe (2004) agrees, stating that, if the intra-forest clearings on hydromorphic soil have a hydrological origin, their maintenance is due to large mammals. In this case, their presence would be the result of current changes such as grazing or trampling of herbaceous plants by large fauna, particularly elephants. Although the factors for the creation and maintenance of clearings on hydromorphic soils are not yet fully understood, it seems that the main factor in their creation is the relatively thin thickness of a waterlogged substrate.

These clearings provide large quantities of herbaceous vegetation for herbivorous animals: studies of the activities of several mammal species have shown that they represent an important trophic space for these animal species, in connection with the abundance of herbaceous plants and the richness of the soil and plants in mineral salts (Magliocca, 2000; Magliocca & Gauthier-Hion, 2001).

Table 7: Comparison of water analysis results of Malambo samples against animal preference

N°	Parameters	Beed and Myers (2000)		Results found			
		Minimum concentration threshold	Extreme concentration Maximum	SE-05	SE-06	SE-0	Average
01	pH	6,8 - 7,5	< 5,5-8,5 >	4,93	5,84	6,51	5,76
02	Phosphorus (P)	0-1 mg/l	-	0,012	0,041	0,052	0,04
03	Sodium (Na)	1000 mg/l	2000	280,6	230,0	4986,4	1832,3
04	Bicarbonate	500 mg/l	-	550	420	520	496,6
05	Calcium (Ca)	500 mg/l	1000	250	1024	1048	774
06	Total hardness	0-180 mg/l	-	32,19	51,51	376,69	153,46
07	Conductivity.	200 us/ cm	400	122	100	432	218
08	Chloride (Cl-)	1500 mg/l	3000	532,5	710	532,5	591,67
09	Sulphate (SO_4^{2-})	500 mg/l	1000	51,07	95,57	64,46	70,37
10	Fluoride (F-)	1 mg/l	2	2,51	2,34	3,40	2,75
11	Nitrate (NO_3^-)	200 mg/l	400	27,4	30,5	31,2	29,7
12	Magnesium (Mg^{2+})	250 mg/l	500	49.2	189,6	265,2	168,0

From the analysis of the results in Table 6, it appears that the first seven parameters analyzed (01-07), coincide with the reference results. Indeed, the *Malambo* water has an average pH of 5.7. A concentration of 0.04 mg/l of phosphorus, 18.32 g/l of sodium, 4.966 g/l of bicarbonate, 0.774 g/l of calcium, 15.346 g/l of total hardness and 218 us/cm of electrical *conductivity*. This high salt concentration is an attractive feature for most animal species that come to drink, but also to lick the soil in specific places where these contents are higher. Awad El (2001) considers that the "extraordinary physico-chemical characteristics of wetlands are at the origin of their richness in biodiversity. According to Dany (2004), the high salinity contained in such water would be an attractive factor to animal species as it has levels considered safe in drinking water for domestic animals (Beed & Myers, 2000). The results in Tables (6) and (7) attest to the fact that the water quality of the *Malambo* is attractive to wildlife.

Table 8: Comparison between physico-chemical parameters of water and soil samples in Malambo

	Paramè (average	Samples analysed	s pH	Electrical Conductivity.	P (mg/l)	K (mg/l)	Na (mg/l)	Ca (mg/l)	Mg(mg/l)
01	Water	3	5,76	218	0,04	59,80	1832,3	774	168
02	Soil	7	5,07	165,86	32,24	2.36	7,08	11,50	3,44

The average levels obtained for the seven different physico-chemical parameters in the water and soil show that the water in the *Malambo is* more concentrated than the soil when comparing the levels of (pH), potassium (K), sodium (Na), calcium (Ca) and magnesium (Mg) (Table 6). Due to specific soil and climatic conditions, the *Malambo* of the Itombwe forest massif are surrounded and sometimes penetrated by grassy formations on hydromorphic soil, generally known as "savannahs", "intra-forest clearings on hydromorphic soil" in reference to Lejoly & Lisowski (1997).

3.5   . Traditional knowledge, biological resources and suis generis protection

In the context of a globalized economy, biological and cultural resources have become veritable products whose ever-increasing market value attracts the covetousness of private groups. It has thus become essential to ensure the legal protection not only of these resources but also of the traditional knowledge resulting from them. The latter is now of considerable importance since it constitutes, in the words of the World Trade Organization, a "valuable global resource", from which it must be deduced that it is "in the general interest of humanity to create conditions favourable to its preservation and to the maintenance of the vitality of the peoples and communities that originate and develop such knowledge.

Indeed, through ancestor cults, rites and initiation ceremonies, the populations and the guardians of the tradition show their common will to preserve the natural, cultural and spiritual riches that constitute and convey the sacred forests, including the *Malambo*. In adopting such practices, the traditional authorities and local populations were fully aware that they were taking essential action for survival. Their major concern is to protect and perpetuate their environment, a vital heritage for the safeguarding of the forest capital whose perception of biodiversity is translated, in the Itombwe forest massif, by the vast semantic field of the word ITOMBWE. Although the term Biodiversity was coined by the West, the local communities had their own term under other names. Indeed, the word *Itombwe* has its origin in Kibembe, it comes from the contraction of 3 words namely (i) "Mitotombokyo" which means droplets of water from a dense forest, it means in a word humidity or wet forest. (ii) "Milemandooo" which means everywhere in the forest there was presence of bats, it means there was increased presence of bats. (iii) "MikelyaNgyoku": Elephant tracks, which means that everywhere in the valleys through the forest, there were traces of elephant passages. By combining these three words "***Itombwe lya Mitotombokyo, Lya milema ndoo, Lya mikela ya ngyoku***" which represents the biological diversity that the Itombwe forest abounds in (Ibucwa, 2008).

Historically, customary law has been the primary source of environmental law in African states. The need to protect nature and preserve the balance of the environment is a constant concern in most traditional African societies. This is justified insofar as man lives in harmony with nature of which he sees himself as one of the elements (Kamto, 1996). This concern is reflected in regulations governing tree cutting, the use of common watercourses, hunting, and land use. Although sometimes relegated to a secondary position by the so-called modern legislation inherited from colonisation, this traditional law is still of great help in the legal protection of the environment in the Itombwe forest massif and in some cases plays the role of supplementary law. It is therefore important to consider how African customary law ensures the protection of forests. African customs 'implicitly' ensure the protection of forests by issuing prescriptions before any sanctions are imposed.

The integration of traditional practices in the process of modern management of protected areas or biodiversity could be an effective and sustainable strategy in the context of grafting tradition with modernity on the one hand, and the cultures of the communities of the region on the other, the perpetuation of these practices. Understanding, adapting to and integrating the worldviews, beliefs, representation systems, needs, interests, traditional practices or meanings of the spaces and resources belonging to the local population are also conditions for the success of the nature conservation project, notes (Boya-Busquet, 2006). Also, the traditional management method is already proving to be unsuitable in relation to current environmental contexts, although cultivation practices still favour the protection of the *Malambo*. It needs to be reinforced by adequate strategies applicable to wetlands of international importance, because the current world trend is, as underlined by Bérard et al. (2005), "to conserve local practices, regulate their use and valorize them". Nature conservation was and still is at the heart of some traditional practices of local communities in Itombwe (Kyale & Maindo, 2017). The scope of the said practices is wide. It touches on species and natural phenomena limiting hunting and/or fishing activity, cultural initiation rites, dietary prohibitions and legendary tales inspired by the richness of fauna and flora, including *Malambo*.

3.6 Sacred sites

The primary purpose of the sacred natural sites is to maintain the cultural identity of the local community in the Itombwe Forest. These sites are real places for the preservation of secrets and age-old practices, such as the sacred sites of Iyela and Ashubikulu (Figure 5). A strong communion exists between the people and the spirits or beings that reside in these places. Offerings and sacrifices are made for the conservation of the sites and their occupants. It is in this logic that the values of cultural and spiritual identity as well as the social organization of the population are closely linked to nature and natural resources in general.

Some sanctions may be imposed directly by the community itself. They are often pronounced by the guarantors of the customs and traditions in the name of the community. In other cases the offender is left to the punishment of the offended deities. Those found guilty of transgressing the traditions may be sentenced to live in quarantine outside the community, so as not to defile it, and are forbidden to approach such people on pain of suffering the same fate.

Immaterial sanctions include sanctions of non-human origin, of supernatural nature, interpreted as such in the context of custom. These sanctions can be lightning strikes, death on the spot, due to lightning for example, or slow: long-lasting illness. Depending on the nature of the offence, repentance may be possible, but if not, the outcome is inevitable. The signs of spiritual punishment, especially for slow punishments sometimes leading to death, will allow the community to know that a person has been struck by a deity and even to guess what offence he or she has committed.

Consequently, all the prohibitions and mystical dissuasions that surround these sites contribute to their conservation. These socio-cultural factors influence the safeguarding of biodiversity (Sow 2003).

The preservation of sacredness is important for the maintenance of socio-cultural values. Moreover, if it is adequately recognized and valued, it would clearly contribute to the management of natural resources.

According to Article 8(j) of the CBD, "each Party to the Convention shall respect, preserve and maintain knowledge, innovations and practices of indigenous and local communities embodying traditional lifestyles relevant for the conservation and sustainable use of biological diversity and promote their wider application with the approval and involvement of the holders of such knowledge, innovations and practices and encourage the equitable sharing of the benefits arising from the utilization of such knowledge, innovations and practices.

Traditional science and knowledge should be used to the fullest extent possible in the conservation and management of sacred natural sites. Integrated management schemes should involve the use of natural and social sciences as well as traditional knowledge.

With regard to traditional ecological knowledge, many custodians of sacred natural sites are repositories of knowledge about the biophysical environment through their role as protectors of sacred species, traditional healers and herbalists, and also as decision makers in the context of the agricultural calendar. While respecting and protecting the intellectual property of indigenous cultures, the sharing of traditional ecological science and knowledge must benefit all stakeholders in the sustainable management and conservation of sites. Traditional knowledge holders, natural and social scientists, and students of the humanities and social sciences should be encouraged to work together in an integrated manner to ensure the sustainable safeguarding of sacred natural sites (Wild and McLeod. 2012).

The establishment of more specific laws and regulations for sacred sites with a view to their formal recognition would strengthen the management mechanisms and institutions of local communities, and protect them from external pressures both within and outside officially recognized PAs.

This article reveals that socio-cultural factors such as traditions, customs, beliefs and taboos related to the maintenance of the *Malambo* are determining elements that influence people's behaviours with regard to biodiversity conservation. Linked to social values and norms, these factors motivate people's decisions, practices and actions. Decision-makers should logically value the cultural practices of communities in biodiversity conservation by preserving the traditions, customs and beliefs of the various cultural groups and by integrating the achievements of traditional legislation into the State's legislative and regulatory framework. A very interesting way of valorizing these achievements is the implementation of environmental education projects in schools and university faculties. During the walking classes and scientific expeditions, for example, emphasis will be placed on the discovery of different cultural practices (sacred places, customary legislation) relating to the conservation and development of biodiversity while involving the local community (involvement of the traditional healer, the manager of the sacred forests, the customary chiefs in the learning of pupils and students) Practical activities in sacred environments will enable students to become familiar with species of traditional importance. Research carried out by the students will be shared with the population through conferences and debates, school or academic theatres and popularisation workshops.

These few developments will have shown how African beliefs and traditions *implicitly* contribute to the protection of forests, which is a real issue in the 21st century. The protection of the beings living in them, in this case trees and animals, is therefore intrinsically linked to the rules governing the behaviour of the social group that uses them. It is therefore important for the authorities to rely on these rules for the rational management of forests. The dilemma, however, is the control of sanctions, which, according to the positive law in force in most African States, is the responsibility of the central state authority.
If it is right to say that customary law *unconsciously* contributes to the protection of the forests, because it is undeniable that the norms of behaviour that are secreted do not have this main objective, it is possible to say that African customary law is 'in spite of itself' a guarantor of the safeguarding of our forests. We believe that under the guidance of customary law, the forestry capital of Itombwe can also take root in the baptismal font of African customary law, which is known to be an unwritten law born of ancestral practices which, by dint of repetition, have acquired the force of law.

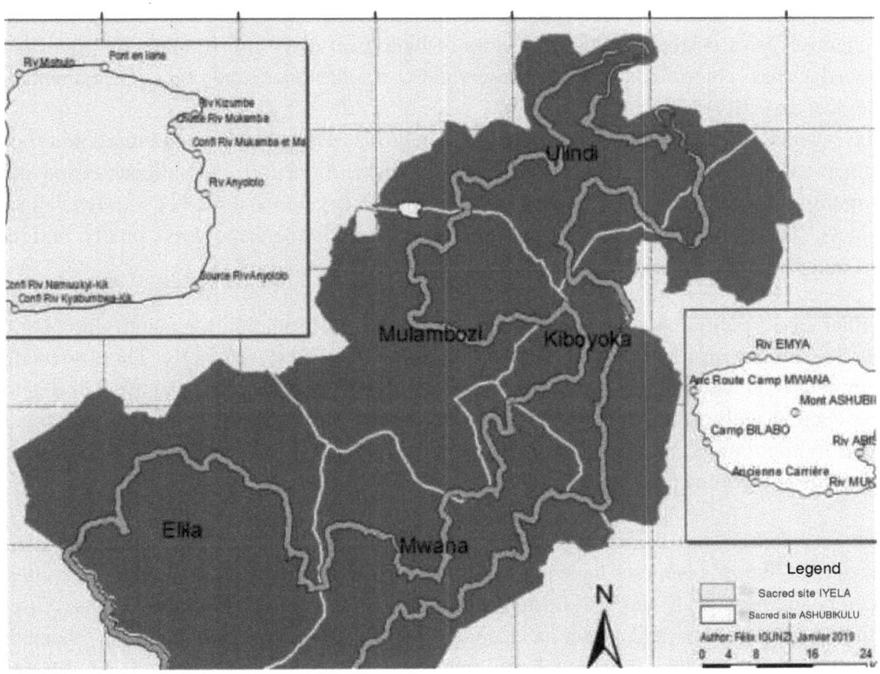

Figure 5. Location of sacred sites in the Basile chiefdom in the Itombwe Nature Reserve.

3.7 Relevance of traditional knowledge and the need to protect traditional knowledge

Indigenous and local communities rely on biological resources for a variety of daily uses, and see themselves as custodians and protectors of biological diversity. In this way, traditional knowledge has contributed to the preservation, maintenance and even enrichment of biological diversity century after century.

Today, genetic resources are being put to multiple uses, both commercial and non-commercial. In many cases, the very properties that made them useful to indigenous and local communities are now being used by economic agents to develop and market widely available products. They are also used by researchers to better understand biodiversity and the intertwined proliferation of terrestrial life forms.

In either case, traditional knowledge is a vital source of information for identifying uses of genetic resources from which humanity as a whole can benefit significantly. This knowledge is particularly valuable to bio-prospectors or users of genetic resources, who use it as a guide in connection with plants, animals or micro-organisms whose useful properties are already known. Without this knowledge, many species currently used for research purposes or for the manufacture of commercial products would never have been identified.

Local and indigenous communities: Over the centuries, the main decision-makers and managers of natural resources have been fairly small and well-defined human communities - gatherers and hunters, fishers and farmers, transhumant and nomadic herders, resource users of forests, watersheds and water sources, builders of terraces and irrigation channels. Biodiversity and cultural diversity have thus evolved together in thousands of different contexts, and 'community governance', too often poorly documented and understood, has developed a wide variety of rules concerning access to and use of natural resources.

Users: The beneficiaries of benefit sharing are in principle the local communities and indigenous peoples who use the forest and/or have traditional knowledge about it. One of the major challenges in setting up and ensuring the effectiveness of a benefit-sharing system is to identify the beneficiaries of this sharing. It is therefore important to understand the occupation of the space and the use of the resource by the Local Communities and Indigenous Peoples (LCPI). In this sense, the right to benefit sharing and the identification of these beneficiaries are closely linked to property and use rights.

Traditional knowledge thus has important implications for access and benefit-sharing of genetic resources. It is essential that traditional knowledge be appropriately valued by those who use it. This includes ensuring that access to traditional knowledge associated with genetic resources is conditional on the prior informed consent of the indigenous and local communities concerned, and that they receive fair and equitable benefits from their use.

The Convention on Biological Diversity (CBD) has established a Working Group on Traditional Knowledge to guide and facilitate discussions among States, indigenous and local communities, and other interested parties regarding traditional knowledge. It provides an opportunity for indigenous and local communities to share their views and recommendations on related issues.

Article 8(j) of the CBD provides for the need for states to respect, preserve and maintain, and promote the wider use of traditional knowledge, with the agreement and participation of the indigenous and local communities concerned.

Thus, when users wish to use traditional knowledge, whether for research or product development, it is their responsibility to obtain the prior informed consent of the indigenous and local communities concerned, and to negotiate mutually agreed terms for the equitable sharing of benefits that may arise from the use of such knowledge.

3.7.1 The CBD as an important instrument for safeguarding traditional knowledge in the Itombwe Forest.

A novel approach to local knowledge Decision VI/10, on article 8(j), captures the philosophy of the Convention. It can be summarized as follows: indigenous and local communities are very dependent on their natural environment and the tangible and intangible resources derived from the components of biodiversity. In turn, these only exist and have been maintained to this day thanks to the sparing (sustainable!) use and efforts and know-how that local human societies have been able to develop over thousands of years. The first consequence of this decision is that the elements of nature and biodiversity must be included in a heritage. If the Convention has renounced to make biodiversity a world heritage, it recognizes its division into a multitude of local heritages. This implies, among other things, the legal recognition of this heritage link and the establishment of adapted access rules (Cornier et al. 2002). The Convention recognizes that local communities have a legitimate right to "their" biodiversity and allows them to control access to it: it recommends, but does not yet require, that interested users ask communities for "prior informed consent.

In any case, the work on local naturalist knowledge is far from being completed. The decisions and texts that emanate from it insist on the fact that we are still at the beginning of a long process.

We conclude this work by expressing the wish that these wetlands be rapidly integrated to make the NRT a RAMSAR site (http://www.ramsar.org/ris/key_ris_index.htm), as they contain a high biodiversity. As such, the *Malambo* will be maintained as sources of biological diversity and by providing the water and primary productivity on which countless species of plants and animals depend for their survival. In addition, they will support high concentrations of birds, mammals, reptiles, amphibians, fish and invertebrates and will also be important repositories of plant genetic material. The interactions between physical, biological and chemical elements such as soils, water, plants and animals, allow the *Malambo,* as part of the "natural infrastructure" of the forest massif, to perform many vital functions, including water storage; drought mitigation groundwater recharge and release; water purification; retention of nutrients, sediments and pollutants; and stabilization of local climatic conditions, particularly rainfall and temperature.

The data collected in this study have only been partially published. Their analysis and publication will continue. The results of this work can be used as a reference for further studies that will be carried out in the future in order to understand the seasonal dynamics of wildlife in the glades and to determine the real impact of landscape changes that may occur in the future.

Conclusion

The clearings and associated sacred forests are a reality in the NRT. They constitute an undeniable space of cultural expression for the local populations. They allow the protection of biodiversity, of which they are today a real sanctuary. This is why they must be safeguarded. Thus, the specific habitats of the *Malambo as* relics of biodiversity in the Itombwe forest massif could hardly have survived without the contribution of the traditional management practices that have contributed considerably to the natural and cultural heritage, creating and conserving landscapes capable of producing multiple goods and services and securing livelihoods. Traditional forest-related knowledge increasingly relies on synergy with modern technology, and reflects a deep understanding of the dynamics of forest ecosystems and the behaviours and characteristics of a wide variety of animal and plant species.

In order for traditional forest-related knowledge to be fully incorporated into sustainable management, the capacity of communities needs to be strengthened and research on traditional knowledge needs to be expanded. Benefit-sharing and other types of joint management arrangements should also be further explored. When disconnected from their natural environment, indigenous communities inevitably lose their traditional knowledge and usually end up being among the poorest people in the world. However, there are some hopeful signs. Forest scientists, for example, are increasingly aware that local communities with traditional forest knowledge can play an important role in helping to implement sustainable forest management. In these modern times, when young people are increasingly distancing themselves from customary knowledge, which is sometimes frightening, the current generation of traditional managers has a heavy responsibility to inform and train. The condition of sustainability of traditional knowledge requires its adaptation to the modern context. This implies that traditional systems must come out of the armoury where they are locked up to open up to other systems, including by supporting and extending the effectiveness of formal administration mechanisms. However, traditional custodians need recognition to strengthen the protection and enhancement of the common heritage of ancestral values. They need to be supported and strengthened. It is imperative that modern legal systems define new modalities taking into account the current realities of African societies.

To this end, collaboration between policymakers, forest managers and local communities is increasingly recognized as a key factor in sustainable forestry. For example, there are many initiatives by indigenous peoples' organizations, non-governmental organizations, national governments and other institutions to safeguard traditional knowledge. In view of the above, collective efforts need to be made and developed between the state and traditional authorities to ensure the conservation and preservation of the sacred forests of the Itombwe Forest Massif, a heritage at risk of being classified as endangered in the face of threats from both artisanal and large-scale mining without any appropriate duty of care policies.

Acknowledgements

We would like to thank the local communities in the different entities of the NRT located in the Mwenga Territory who collaborated closely with the ICCN and WWF teams to carry out this work. Also, this work would not have been possible without the funding graciously provided by the World Wide Fund for Nature (WWF) for biodiversity conservation in the NRT. We are very grateful for this.

Bibliographic references

AUBREVILLE, A. (1948). Richesses et Misères des Forêts de l'Afrique Noire Française. Paris.

AUBREVILLE, A. (1951). The concept of association in the dense equatorial forest of the lower Ivory Coast. Bull. Soc. Bot. France. Mémoire. 145-158.

AWAD EL, S. (2001). French methods for wetland designation for Ramsar inventory: Lessons learned and proposals for improving wetland management in Egypt. Mémoire DEA. Université Senghor.92p.

BEED, D & MYERS, H. (2000). Water, an essential nutrient, 24th Dairy Cattle Symposium. CRAAQ, Quebec, Canada, pp. 71-91. Saint-Hyacinthe, Quebec,

BERARD L. & P. MARCHENAY. (2004). Les produits de terroirs, entre cultures et règlements, CNRS Editions, 232p.

BERARD L., CEGARRA, M., DJAMA, M., LOUAFI, S., MARCHENAY, P., ROUSSEL, B. & VERDEAUX, F. (2005). "Savoirs et savoir-faire naturalistes locaux: l'originalité française", VertigO - la revue électronique en sciences de l'environnement [en ligne], Volume 6 Issue 1/ May 2016, online 1 May 2005, accessed 22 March 2018. URL: http://vertigo.revues.org/2887; DOI:10.4000/vertigo.2887.

BOUPOYA A. (2010). Flora and vegetation of inter-forest clearings on hydromorphic soil in Ivindo National Park (North-East Gabon). Doctoral thesis, Université Libre de Bruxelles (France), 246 p.

BOYABUSQUET, B. (2006). "Sustainable integrated strategies: traditional ecological knowledge and adaptive management of spaces and resources", VertigO - the electronic journal in environmental sciences [Online], Vol. 7, No. 2, URL: http://vertigo.

BRAUN-BLANQUET J. (1932). Plant sociology: The study of plant communities. Mc Graw Hill, New York and London, 439 pp.

CORMIER-SALEM M.-C., D. JUHE BEAULATON, J.-B. BOUTRAIS & B. ROUSSEL, 2002. Patrimonializing tropical nature. Local dynamics, international issues. Colloques et Séminaires, IRD, Paris, 467 p.

DANY, C. M. (2004). Water: nutrition and food. MAPAQ/Direction de l'innovation scientifique et technologique.14p. Retrieved January 17, 2017, from http://www.agr.gouv.qc.ca.

DEFAILLY, D. (2010). Itombwe Forest: Socio-economic issues and nature conservation in the Congolese context. 26p.

DEVRED, R.(1961). New mechanographic method of phytosociological investigation in equatorial forests Vegetatio, **10** (1): 57-66.

DOUMENGE, C. (1998). Forest diversity, distribution and dynamics in the Itombwe Mountains, South-Kivu, Congo Democratic Republic. Mountain Research and Development, 18, 249-264.

DUCROS, L. B. (1998). Anthropology and the Environment, The Environmental Issue in the Social Sciences. Eléments pour un bilan. PEVS CNRS, Lettre PIREVS n°17, p 13-23.

DUVIGNEAUD P. (1949). Les savanes du bas Congo. Essai de phytosociologie topographique. Lejeunia, **10**: 1-192.

GOVERNORATE OF R\PROVINCE OF SOUTH KIVU. (2016). Arrêté provincial portant mesures provisoires des limites issues de la délimitation participative de la Réserve Naturelle d'Itombwe.

HART, J. & MUBALAMA, L. (2005). Conservation of gorillas and chimpanzees in the Itombwe Massif. In Gorilla Journal of Berggorilla & Regenwald Direkthilfe No. 30, pp 7-8

IBUCWA, J.P. (2008). Study of strategies for reconciling conservation with education: The case of the Itombwe forest massif, 2008. 78p.

ICCN-WWF, 2018. Map of the spatial distribution of "Malambo" in Itombwe Nature Reserve.

ICCN/WWF. 2018. Spatial distribution map of specific glade habitats known as *Malambo* in Itombwe Nature Reserve.

IGUNZI, F. (2017). Ecological and socio-cultural potentials of the "Malambo": Strategy for the designation of the Itombwe Nature Reserve as a Ramsar site. South Kivu Province, DR Congo. Master's thesis in Development from Senghor University, Alexandria, Egypt.

JEUNE AFRIQUE. (1978). Les atlas Jeune Afrique: République du Zaïre. Jeune Afrique, Paris: 72 p.

JEUNE AFRIQUE. (1993). L'atlas Jeune Afrique du continent africain. Groupe Jeune Afrique & Ed. Jaguar, Paris: 175 p.

KAMTO, M. 1996. *Droit de l'environnement en Afrique*, EDICEF/AUPELF, p. 66.

KYALE, K. J & MAINDO, M. A. (2017). Traditional Nature Conservation Practices Fact Checked Among Riverine Peoples of Yangambi Biosphere Reserve. European Scientific

Journal March 2017 edition vol.13, No.8 ISSN: 1857 - 7881 (Print) e - ISSN 1857- 7431.p. 328-356

LEBRUN, J. (1947). Vegetation of the alluvial plain south of Lake Edward. Exp. Parcs Nat. St. Albert. Lebrun Mission (1937-1938). Inst. Parcs Nat. Congo belge, I, 467 p., II: 471800.

LARRERE, C. (1997). Du bon usage de la nature: Pour une philosophie de l'environnement, Paris.DOI : 10.1016/S1240-1307(97)81559-2. Paris.

LEJOLY J. & S. LISOWSKI. (1997). The vegetation of clearings on hydromorphic soil in the Odzala National Park (Congo-Brazzaville). Colloques phytosociologiques, XXVII, 371-382.

LEONARD J. (1950). Botany of the Belgian Congo. Plant groups. Encyclopedia of the Belgian Congo. I: 345-389.

LEONARD J. (1952). Preliminary overview of pioneer plant groups in the Yangambi region (Belgian Congo). Vegetatio, 3: 279-297.

LETOUZEY R. (1982). Manual of forest botany. Tropical Africa. Volume 2A. CTFT, Nogent-sur-Marnes, 210 p.

LETOUZEY R. (1983) Manuel de botanique forestière. Tropical Africa. Tome 2B. CTFT, Nogent-sur-Marnes, 461 p.

LETOUZEY R., (1985). Explanatory note for the 1:500 000 phytogeographic map of Cameroon. Encyclopédie biologique 69, Edition Paul Le Chevalier, Paris 511 p.

LUKETA-SHIMBI, H. (2005). Sacred forests and conservation: the case of the Democratic Republic of Congo. Unpublished paper.

MAGLIOCCA F. (2000). Etude d'un peuplement de grands mammifères forestiers tropicaux fréquentant une clearing : structure des populations ; utilisation des ressources ; coexistence intra- et inter-populationnelle. Doctoral thesis, University of Rennes 1, France: 285 p.

MAGLIOCCA F. & GAUTIER-HION, A. (2001). Glades in tropical forests: Areas to be protected as a priority. Canopy, 20: 1-9.

MAHAMANE A. (2005). Floristic, phytosociological and phytogeographical study of the vegetation of the W Regional Park of Niger. Thesis of doctorate. Université Libre de Bruxelles. 517 p.

MINISTRY, ECN EF. (2006). Order No. 038/CAB/MIN/ECN-EF/ of 11 October 2006 establishing the Itombwe Nature Reserve.

MUBALAMA, L., MATABARO. A. & HAMULONGE, J. (2013). The joint framework approach as a strategy for participatory delineation of the Itombwe and Bushema forest massif. In Mwapu, I. P., Karhagomba, B. I., Mapatano, S. & Niyonkura, D. Governance of Collective Natural Resources of Fragile Ecosystems in the Great Lakes Region, Democratic Republic of Congo. Les Editions du CERUKI pp.246-258.

MUBALAMA, L. ET AL. (2017). On the Road to extinction? A population estimate of Great apes in Itombwe. In Gorilla Journal of Berggorilla & Regenwald Direkthilfe. 54:5- 13.

MUBALAMA, L. (2018). Status and management of protected areas. Les Editions de l'Université Officielle de Bukavu.320 pp.

MUHIGWA, J.B., BIREMBANO, R., MUCHUKIWA, B., MUGANGU, S., BANTU, J.M., AND USUNGO, J. (2011). Itombwe Massif - The People and the Land. Bureau d'Études Scientifiques et Techniques. Bukavu, Democratic Republic of Congo.

NGANONGO J.B. (2000). Suivi des salines; Odzala National Park, Congo. Monthly report January.

NOUPA P. & NKONGMENECK B.A. (2008). Influence of forest clearings on the spatial distribution of large mammals in the dense forest of the Congo Basin: the case of Boumba-Bek National Park (South-East Cameroon). International Journal of Biological and Chemical Sciences, 2 (2): 185 - 195.

OMARI, O. (1999). The Itombwe Massif, Democratic Republic of Congo: Biolocical Survey and conservation with an emphasis on Grauer's Gorilla and birds endemic to the Albertine rift. Oryx 33,301-931.

UNITED NATIONS. (1992). THE CONVENTION ON BIOLOGICAL DIVERSITY (CBD). Earth Summit in Rio de Janeiro- Brazil.

PARMENTIER, I. & MÜLLER, J. V. (2006). Grasslands and herbaceous fringes on inselbergs in Atlantic central Africa. Phytocoenolologia, 36 (4), 565-59.

PRITCHARD, D. (2010). Ramsar handbooks for the wise use of wetlands, Handbooks 17. Ramsar site designation.132P. (4 ed.).

RAUNKIAER, C. (1934). The life forms of plants and statistical plant geography. Oxford University press, 632 pp.

ROUE, M. (2012). "History and Epistemology of Local and Indigenous Knowledge", Journal of Ethnoecology [Online], 1 | 2012, Published online 02 December 2012, Accessed 30 September 2016. URL: http://ethnoecologie.revues.org/813; DOI: 10.4000/ethnoec.

ROUSSEL, B. (2005). "Local knowledge and biodiversity conservation: strengthening community representation." Movement. vol 4, no 41. pp 2-88. URI: www.cairn.info/revue-movements- 2005-4-page-82.htm.

SENE C., D. P. (2013). "Traditional practices and sustainable management of natural resources: A case study of coastal and marine sacred natural sites in Jaol-Fadiouth", ASRDLF - Call for papers - Style sheet, 18 p.

SENTERRE, B. (2005). Recherches méthodologiques pour la typologie de la végétation et la phytogéographie des forêts denses d'Afrique tropicale. PhD thesis, Université Libre de Bruxelles, 354 p.

SOW M. (2003). Rôle *des structures traditionnelles dans la valorisation de la biodiversité, In* Butare I. *Pratiques culturelles et conservation de la biodiversité en Afrique de l'Ouest et du Centre.* IDRC, Zoom Edition: 205-212.

TROUPIN G. (1966). Phytosociological study of the Akagera National Park and Eastern Rwanda. Thesis for the agrégation de l'enseignement supérieur, Université de Liège. 293 p.

THIBAULT, M. (1991). La gestion des populations naturelles de truite commune en France analysée dans une perspective historique (1669-1986), Maisse G. (ssdir.), La Truite, biologie et écologie, Ed. INRA, p. 239-293. Baglinière JL.

TUBIANA, L. (2005). "Local naturalist knowledge, a global issue", in Bérard L., Cegarra M., Djama, Louafi S., Marchenay P., Roussel B. &Verdeaux F. (Eds.) (2017). Biodiversity and local naturalist knowledge in France, CIRAD, IDDRI, IFB, INR.

VANDE WEGHE J. P. (2004). Forests of Central Africa: Nature and Man. Lannoo, Tielt Belgium, 367 pp.

VANDE W J. P. (2006). Ivindo and Mwagna. Blackwater, rainforest and Baïs. Wildlife Conservation Society, Libreville (Gabon), 272 p.

VANLEEUWE H., CAJANI S. & GAUTIER-HION A. (1998). Large mammals at forest clearings in the Odzala National Park, Congo. Revue d'Écologie (Terre Vie), 53: 171-180.

WHITE, G. (1983). The vegetation of Africa, a descriptive memoir to accompany the UNESCO-AETFAT-UNSO vegetation map of Africa. Natural Resources Research, UNESCO, 20, 1-356.

WILD, R and McLEOD, C. (Editors) (2012). *Sacred natural sites: Guidelines for protected area managers.* Gland, Switzerland: IUCN. xii + 108pp.

WILSON, C. J. (1990). A preliminary survey of the forest of the Itombwe Mountains and the Kahuzi-Biega National Park extension, East Zaire: July-September 1989.

WWF/ICCN. Itombwe Program. 2018. Report on the documentation and demarcation of the external boundaries of Itombwe Nature Reserve (BASILE and WAMUZIMU chiefdoms): Phase 1.

Printed by Books on Demand GmbH, Norderstedt / Germany